ANGELS WATCHING OVER ME

The Memoirs of Gilles Goddard

GILLES A. GODDARD
TRACY KRAUSS

Angels Watching Over Me – The Memoirs of Gilles Goddard

Copyright © Gilles Goddard, Tracy Krauss, 2020

All rights reserved.

No part of this book may be reproduced in any form or by any electronic or mechanical means, including information storage and retrieval systems, without written permission from the author, except for the use of brief quotations in a book review.

ISBN: 978-1-988447-62-9 (paperback)

ISBN: 978-1-988447-52-0 (kindle version)

ISBN: 978-1-988447-51-3 (epub)

Published by Fictitious Ink Publishing

Tumbler Ridge, BC, Canada

V0C 2W0

Scripture quotations are from the ESV® Bible (The Holy Bible, English Standard Version®), copyright © 2001 by Crossway Bibles, a publishing ministry of Good News Publishers. Used by permission. All rights reserved.

FOREWORD

I met Gilles in 2008 when we started attending the same church. Even then, he was a hardworking man of strong faith who could spin an interesting yarn. In late 2019, he asked if I could help him redeem an airplane ticket he'd received as a gift. He wanted to use it visit his wife in the nursing home where she resided, but wasn't sure how to do it. After hearing the story behind it, I wrote an article for our local newspaper called *"A Valentine For Gilles"*. I mentioned to him that he should start recording his life story, since it was so fascinating. Soon after, he called me up and asked, "Were you serious about my story?" And thus, this project began. Much of what you will find here is an uncensored glimpse into the life of an ordinary man who has lived through some extraordinary - and sometimes cruel - circumstances. Through it all, he has maintained a simple faith in God. I hope you will be as inspired by Gilles' story as I have been.

Tracy Krauss

***

For He will command His angels concerning you to guard you in all your ways.

Psalm 91: 11 (ESV)

Chapter One

A TRUCKER'S LIFE

I drove truck for fifty years on the highways of North America, mostly between the USA and Canada, with a couple of little slips into Mexico to load up produce. I calculate that to be about five million miles, or 8 million kilometres. I've been to every State except Hawaii and Alaska. I've been to the Yukon border and all across Canada. Sometimes I think I'd like to retrace my steps and visit all the places I went, but I'd have to live to be 130 years old, I think, in order to do that! I hauled anything and everything. Propane, lumber, equipment, sea cans, steel - you name it.

In all that time, God has been with me all the way. I used to have a license plate on the front of my truck that said, "God is my co-pilot." And I believe it's true. I've seen a lot of things in my life. Been through a lot of tragedies, suffered a lot of heart-ache, had a lot of adventures - and a few near misses, too. I kept God so busy, I don't know how He's had time for anyone else!

I remember one winter day in Pennsylvania – and Pennsylvania is, for all intents and purposes, in the mountains. You're either going uphill, downhill, or around a sharp corner. It was late in the afternoon, and it hadn't rained down in the lower country, but up in the hills it had, and it was freezing rain.

I was coming back into Bayonne, New Jersey, with an empty tanker. I came to this hill and there was something at the bottom that I needed to steer clear of. When you're trying to slow down on an icy stretch of road, sometimes it can be difficult, so I did the best I could. At one point, my tractor was forty-five degrees to my trailer. When I finally got it stopped, that's the way I was sitting.

Then this other vehicle came along. They asked if I was hurt or if there was any damage. I said no to both. There was just this one muffler guard that got bent a little bit because I'd cranked so sharp, but the truck itself was fine.

We looked at the situation and decided it was best if I could try and get straightened around before somebody else came along. So, I started back and forth, crimpin' and crampin' until I got the rig straightened enough to get going downhill again. I got about halfway down that hill and the ice turned to water because of the temperature change. After that, it was smooth sailing.

In that incident, I believe God was with me. Nobody got hurt, and nobody got creamed out. The damage was nothing - just the little shield over the muffler.

At the time, it seemed like a minor incident, but all these little things add up. Over my lifetime, I've had my share of both big and little events, but my relationship to God has been just as important during the little ones as it has been with the big, because I know I can depend on Him. Jesus said, "I will never leave you. You will never be alone."

In all my years, I got into the habit of saying a little thank you prayer when I get up. One of the first things I say is,

"Good morning, Holy Spirit." I also say good morning to the angels guarding me. I say good morning to Jesus, and good morning Father God. I don't leave anybody out! I ask God to bless my wife and help me to be a blessing to someone else. That's a good way to start the day, I think!

You can't change yesterday, and there's no point worrying about tomorrow. It's not where you come from, but where you're going that counts. But, I do believe we can learn from the past. Sometimes, God allows us to go through things so that we can learn a lesson ourselves, or help someone else along the way. I'm so grateful for all I've been through and can only pray that someone, somewhere, will benefit from my story.

I believe we're all here for a reason. God has work for us to do, and until that gets done, we're gonna be here until He decides we don't belong anymore. Satan probably wants to get rid of us right now because we're doing too much damage to his outfit. But he can't take our life from us. If God is for us, who can be against us?

I'm here - and I shouldn't be! I should have been dead forty-five or fifty years ago, but God must have something for me to do yet, 'cause I'm sill kickin' - and I'll be here until it gets done! That's what keeps me going.

As long as we have God to think about and thank for our life, we'll be alright.

Chapter Two
EARLY YEARS

WHERE IT ALL BEGAN

I was born by a little town called Island Brook, Quebec. The area is called the Eastern Townships. Our place was about 125 miles east and a little bit south of Montreal, as the old road used to go. Now, with the new highway, it's probably only about 90 miles.

I grew up there until we moved into Sherbrooke, Quebec, when I was six or seven years old. My parents split up - not very common in those days - and so my mom was a single parent most of the time. I had one sister - she was the oldest - and one brother. My sister was two years older than my brother, and my brother was two years older than me. I was the baby.

I got along with my sister, but for some reason, my brother and I never did get along very well. We managed to make up in our latter years, but now he has dementia. He still has a fair amount of memory, but he can't really carry on a conversation or anything and doesn't drive. His boys take him shopping and stuff.

Anyway, it was hard for me, sometimes. My mother was a Christian, but she couldn't stand my father, so they couldn't live together. I spent time with my mother, and I spent time with my father. My dad and I got along really good. My Dad was the one who delivered me.

The neighbour woman had said, "Frank, when the time comes, you call me. Just open the window and holler, and we'll hear you." They were close neighbours, obviously. It was September 15 and in the middle of the night, and at that time of year, it could be nasty weather. So, by the time Mrs. Kerr got there, I was already born.

Nobody ever mentioned this next part to me except my Dad, but he said my Mom tried to squeeze me when I was being born so that I wouldn't come out alive. Apparently, she didn't want me. Now, my Mom being a Christian, to do something like that, the enemy had to be working. You can't just say, "She was bad," because you never knew her. Maybe she couldn't face another child, especially if she was unhappy in her marriage. In any case, it seems like a rough way to start life, but I guess right from the beginning, God had a plan for me.

∼

Childhood Mishaps

My schoolteacher had this farm just up a ways from the school. Island Brook was a small town - you could count the number of houses on one hand. His barn was a gable roof, going down both sides. He and two or three other guys used to go inside the doorway and shoot the pigeons off the roof of the barn to see how good of a shot they were.

Well, I was just a little snotty nosed kid, about five or six, fooling around in his garage. I climbed up on the roof overhead and I fell through, landing on the hard floor. Everyone was shocked. "What did he break? Is he dead?" and all manner of fuss. But, I was okay, and I got up laughing! Not a broken bone.

Around that same time, I was out working with my Dad one day. He was picking rocks in a field for this same guy. In those days, people didn't have all the fancy machinery that we have today. Then, you had a horse pulling a wagon, and you picked those rocks up and put them in the wagon. That's how they built their fences and so on, out of these rocks. There was this one big flat rock, and I thought, "I'm going to show my Dad how strong I am." I had to bend over and get my hands around it, and I gave a huge heave. The rock let go, and I followed it down. My head landed on a sharp edge, and it split my head open.

It must have been pretty bad. Dad left the horse and the wagon and carried me to the house. In both those two instances, I didn't even cry. Crying was not the thing to do!

What I see, now, though, is God continually protecting me. Besides being a naughty boy, I could have died from that incident - from any one of them, or been insane with brain damage. But God was watching over me right from a young age.

I had lots of other mishaps as a boy. For instance, both of my knees have huge scars.

The one on the right is a square with two corners that come together like a 'V'. (I like to think it's V for victory!) That was from my uncle's motorcycle. We had gone swimming down by the river and he had parked his bike on the right side of the road. There was a fence and then some bush, so my uncle went behind the bush to change into his swimming trunks. When we were done swimming, he went back behind the bush to change back into his clothes. I had knee hoppers on, so I never needed to change. Instead, I was on his bike pretending to drive it, turning the wheel and everything. After he was changed, he said, "Okay, you'll have to go with your mother now, cause I need my bike."

I had the wheel turned, and then I hopped off on the wrong side, and the bike started coming over. My uncle took one jump, landed on the right side of the fence, and grabbed that bike! The spark plug had already gone right into my knee, but he'd stopped it enough that it never broke any bones or anything.

Someone picked me up, put me on my mother's lap, wrapped a towel over my knee, and said, "We're going to go see the doctor!" We made it to town - it probably took half an hour to get there. It was a Sunday afternoon, but the doctor fixed me all up. He used three clamps on each side and then told me I had to stay in bed for a week. I thought, "How am I going to stay in bed for a whole week?" I was just a kid!

But, boy, did I ever have friends! They all wanted to come and see the big scar! I was famous for a couple of weeks!

I injured the other knee on a Halloween night. My brother, our friend and I were all riding our bikes, but the chain had broken on my bike. So, my brother tied a rope from the handlebars of my bike to the seat on his bike, and he was towing me around. It was a fairly long rope, and we were doing alright. We were biking on a long gravel road, and there was rain and snow, and suddenly he hammered on his brakes.

I had no brakes because I had no chain. The rope got tangled in his bike, and I kept going over the handlebars and landed on the road - right on a sharp rock that got embedded in my knee. Talk about, "Ouch!"

They got me home, and of course, mother had to take out a bunch of dirt and gravel. I still have the scar to prove it - right below the knee cap, while the other one is above the knee cap.

It could have been a whole lot worse, but God was protecting me, and it was also a good lesson: Don't ever ride a bike with a rope tied to the handlebars!

Childhood Triumphs

You may be familiar with the Lancaster four engine bomber that Canada used during World War II. It was quite a popular airplane. It's a British plane, and it did more to win the war than any other plane used. There was one at Napanne, Ontario. I was an air cadet, and we went there for a visit. I was only thirteen years old, but I got to go for a ride. It was a thrill to get to fly in a Lancaster Bomber, I can tell you.

I was always quite athletic. My high school athletic coordinator was very enthusiastic, especially about skiing, and encouraged me to do a lot of skiing in school. We went to competitions, and there is good skiing in Quebec. My friend Gerry and I were at a competition this one time, and he asked me if he could use my smaller, lighter skis in the cross-country event. He was a few years older, kind of like a big brother to me. He won the cross-country event, and because of it, he had the most overall points at the competition! I also accumulated a lot of points and came in second overall. It was fun, but it made me think how important it is not to let it become an obsession. Sometimes I wonder if I could have pursued sports more, but my mom was a single mother, working as a waitress and raising three kids. Reality.

When I was thirteen I worked as a golf caddy. I got $1.50 for going around carrying someone's bags. One time I earned $5.00! Afterwards, I went to the restaurant where my mom worked as the head waitress, and I had a huge smile on my face. I gave my mom that five dollars, and I got the biggest banana split sundae!

I also had the record at the YMCA in Sherbrooke between all the high school kids for swimming the longest distance in the pool under water. Most could go two lengths, but I could go three lengths. Some might say I'm long winded! But I always liked doing things that kept me in shape.

It came in handy at times. I was working one summer in a tobacco field in southern Ontario. A bunch of us used to go there from Quebec and work during the tobacco harvest. That summer we had a really good crew. Some crews worked until six or seven at night, but we were always done by three o'clock in the afternoon. What do you suppose we did with the rest of our time on a nice warm summer day by Lake Erie? We went swimming!

Anyway, a bunch of us went swimming one day, and this one guy - from Newfoundland, I think - was in the water when a little breeze came up, and the waves started rolling in. I heard him start to scream. I looked around, and there he was, his arms waving and him screaming. I thought, "Oh, oh! This guy's not much of a swimmer..." So, I swam out to him.

Now, there's a way of approaching a drowning person without putting yourself in danger of them grabbing you the wrong way and pulling you down, too. This I learned through the YMCA and Red Cross, because I had all of the safety training and badges they had at the time. I could have been a lifeguard. I went down under him, and came up behind him with my arm around his chest. Both his arms came up and grabbed onto my arm, and I thought, "Good! I've got you!" We bobbed up, and I told him to settle down, and got him calmed down, and we managed to get back to shore.

Later, this guy broadcast all over the place how I had saved his life. And I suppose I did, too, but I just did what anybody else would do. What I had been trained to do. I happened to be the guy who was there, so I didn't make a big deal of it, but to him it was a big deal!

I didn't ever want to be anybody important. I was just another farm kid. But God puts you places. Puts you in situations.

∼

Dad

My Dad was illiterate. He couldn't read or write, but he learned to write his name for signing. That's it. He never had the chance to go to school, because he started following his dad around the traplines when he was nine years old. He forgot more about trapping than I ever learned! It was a different time, before modernization. School wasn't seen as necessary if you weren't going to need it to make your living.

He became a cook for the CN Railroad, and he also used to cook for some lumber camps. Somebody asked him one day, "Frank, how can you know what you're buying if you can't read the label?"

He said, "Well, strange as it may seem, my nose still works!" There's always pictures on labels, and he had a very good sense of smell. He cooked all kinds of stuff. He was a good cook.

I was with him one winter in the bush when he was cooking for some friends of ours who owned a lumber camp.

There were six or eight lumber jacks out there, and Dad would go out at four or five in the morning to see if he could shoot a deer or something for food. In those days it was so cold, with so much snow that the deer would walk along the paths that the lumber jacks had made, and they would chew on the trees that they had felled.

One day Dad shot a deer, but it got away, wounded. He came back to camp and started getting breakfast, but told me to go out and follow that deer to try to bring him back. I followed him most of the day, but I couldn't find him. At suppertime when I got back with nothing to show for it, the boys had a good laugh. I was probably fourteen or so.

Dad loved to hear God's word, though. I used to read to him quite a bit, especially when we were alone. He and I didn't always understand each other, but he enjoyed listening to the Bible, since he couldn't read it for himself. Especially the gospels.

So, one night, when we were in the trailer and I was reading, there was a knock at the door. It was around ten o'clock and it was about 50 below zero. I looked at him and he looked at me... We wondered who could be knocking at that time of night in the middle of winter.

I opened the door, and this man stood there, dressed in a black leather coat right down to the ground. He had to be six foot three or four. I asked him in, and we shut the door. He never gave a name or anything, but we talked for a few minutes, and then he just left. Dad and I didn't know what to think!

In the morning, when we went outside, there were no tracks. Dad said, "It had to be an angel." I said, "You might be right!" I'd never seen such a man with such a long black coat and everything! To me, it was a visitation, just letting us know that we were being watched and protected.

One day, Dad told me about a conversation he'd had with God. He also worked skidding logs, and his horse stepped into a hole and hurt his leg. The animal was limping really badly. Dad couldn't afford a new horse, and he didn't want to shoot this one, so he sat down on a and stump prayed. "God if you're real - if you are who my boy says you are - then if you can heal my horse, I'll believe everything my boy has been reading to me, and accept Jesus as my Saviour."

Apparently, he stayed there two minutes, got up, and led his horse to the barn. He took the harness off, fed him, watered him and everything. The next day, he saddled up the horse, and it was just like brand new! That's how my dad became a Christian.

Someone said once, "He became a Christian over a horse?" Well, it could have been a pig or a cow or a dog! It just happened to be a horse. His was a simple faith.

Chapter Three
IN THE NAVY

YOU'RE IN THE NAVY NOW!

I joined the navy at seventeen. My mother had to sign for me, giving me permission to join.

We were living in Island Brook at the time, and I took a bus the 135 miles to Montreal to listen to their "spiel" at the enlisting office. Afterward, the recruiting officer asked, "How old are you?" When I told him I was only seventeen, he said, "You need permission." So, I went all the way home with the piece of paper, got my mother to sign it, and went back to Montreal and signed up.

I wound up just outside of Halifax, Nova Scotia. Now, navy boot camp in those days was six months long. Because I was "naval air", I went to Halifax for my courses after boot camp. When I was ready to be drafted onto the boat, our aircraft carrier had just come back from Korea.

This was the early '50s, and nobody knew the future, so when we went out to sea, we practiced a lot of war games. The big practice we did in '54 was what they called "Exercise Mariner". It involved Canadians, British and Americans.

One day, clouds came in from out of nowhere, and we couldn't see anything. Radar in those days wasn't all that good, and we started getting this call from an aircraft. The pilot was looking for a place to land because he was running out of gas. We got a warning from the radio operator on our ship who said the plane sounded really close. They told him to just look for a hole in the clouds.

Sure enough, he found a hole in the clouds and saw our ship! He didn't know who we were - it could have been friend or foe - but he saw that it was a place to land. Once he'd landed, he was some happy to see that we were Canadians! He was American. He spent the rest of the night on our ship, and we fuelled him up and everything. In the morning, he was surprised to see the name of our ship plastered all over his airplane! We'd used a stencil that we had. When he landed on his own ship, they knew where he'd come from, that's for sure!

We also had some hard times - times where people got hurt. It was not anyone's fault, just the nature of the job. For instance, during a war, you'd be required to conduct flying exercises at night, so we had to practice like that, too. You wouldn't be showing any lights in a war, so that's how we trained. Some guys walked off the edge of the flight deck because they couldn't see. They'd either land in a net on the side, or land into a gun sponson. You have to hit pretty hard if you're going to damage six inch steel! Usually, it's the bones that cave in first.

The worst we had was a guy who walked right into the propeller of a plane that was behind him. All there was, was a splash on the windshield. The pilot couldn't fly for a few days after that - it took him a few days to get over it.

All the aircraft on board were propeller driven. We didn't have any jets on "Maggie". When a plane came in, they'd either go up on the left or right side – port, starboard, centre – directed by the deckhands. The plane would park, and there was always a couple of deckhands who would come and tie them down so that they wouldn't move. We had straps and deck bolts and tie downs. So, this deckhand did the tail, and he backed out from under the tail and turned – right into the propeller of the other plane. So, there were things that happened, even during peace. You don't have to be on the front line shooting at each other in order to get hurt or to be affected by tragedy. It made you pay attention, for sure.

I had a particularly close call myself, one time. We were doing exercises, and the planes were all in position to either fly straight up, or go on the catapult. There were two deckhands for each plane, one on each wheel on each side with chalks. You couldn't talk to anybody because of the noise of the engines. All those engines turning also created quite a wind velocity on the deck – anywhere from 100 to 150 miles per hour. So, when you picked up those chalks, you should be able to turn away and grab onto these holes with anchors in them that were used to tie the planes down. You grabbed onto that, and then you lay down on the chalks until the plane moved.

I did that, except there was no hole there! There was nothing! They discovered later that the plane had been parked too far back, too close to the back end. So, when I laid down, the wind got a hold of me, and I started sliding back. I was hoping the aircraft would hurry up and move, but I kept sliding... slow and slow and slow, but still, going steady.

I figured I was going to have to jump once I reached the back end. The flight deck was seventy feet off the water, but I knew I couldn't let myself go straight off the back because those twin props would suck me under, and I'd never make it. When I got about eight to ten feet from the edge, I decided I was going to stand up and jump as far away from the edge as I could to get away from that prop wash.

Just as I was about to do that, I stopped sliding. Nothing happened to the wind - it was still going strong, but I just quit sliding. I looked around, and nobody was grabbing me or anything, but I just wasn't sliding anymore. Then, all the planes moved, and the wind died down. Of course, there were things to do, and it slipped my mind. But I started thinking about the whole incident later, and I came to the conclusion that something abnormal must have happened.

But, that's not the end of the story. When we were at sea, we used to get mail every week, or sometimes every two weeks, depending on where we were. With our aircraft carrier, the captain knew where we were going to be in advance and would send a plane out to get our mail. It was about a week or so after that incident that I got a letter.

There were two people who used to write me - my mother and my pastor. So, I got this letter from my mom. Most of the letter contained nothing special, but at the bottom she put a P.S. She asked if on a certain day, at a certain time, did anything strange happen? I thought that was a funny question. Then I started thinking about it, and that day lined up with what had happened the day I almost went over the back end. I had to ask someone else what time it would have been in the location where it happened against the time it would have been at home. It checked out!

Then, she put a P.P.S on the bottom of the letter. She said, "On that day, your brother got a burden to pray for your safety. He threw himself on his bed, facedown, and started to pray."

Do the Math! It turns out it was at exactly the time that I was sliding toward the edge of the deck and then stopped sliding! At the time, I thought I felt something strange happening, and now I believe it was an angel who stepped in front of me. "You ain't goin' no further, bud!" He stopped me from sliding because of my brother's prayer.

That's where the fact that there is no distance in prayer really hit me. Time isn't even a factor. I mean, God created the universe, so what's our tiny little planet? That's not a far distance for God! His angels were doing double duty for me that day.

~

Number One Fire Suit

As part of our training, we worked at Shearwater airport for a time, which is outside the Halifax naval station. Besides our own navy aircraft, we took in a couple of TWA airplanes that used to use our landing strip. (TWA stands for Trans World Airlines.) This was back in the early '50s, and we didn't have the electronics that we have today. At that time, the air strips didn't have electric lights. They used kerosene pots that we had to put out every evening full of kerosene so they'd burn all night. We'd put them all along the airstrip so that there was light for night landing. In the morning, you'd have to put out the fire and bring the pots back in for refilling. Times have changed.

This duty would last for a month. It gave you something to do because, sitting there for a month, time could go pretty slow. We also had to be part of the 'crash crew'. Our bunks were in a little hut with the fire engine, and sometimes, they'd do a test. The bell could go off at two in the morning. The driver slept in the truck, so all he had to do was wake up, sit up, start the truck, and go. The rest of us had to make sure we were on the truck with him. It didn't matter - snow or whatever – you had to be ready. It was part of special duties training.

This was before I got drafted on board the aircraft carrier. It was necessary to be familiar with what to do if a fire occurred or a crash happened on board the ship. Being the smallest guy on the crew, I was number one fire suit, because I was the only one who could wedge into the security exit – the emergency exit, basically – to haul out the pilot or the navigator or whoever was caught after the crash. So, I found that quite interesting.

We learned a lot about different types of fires. To put out an oil fire, you smother it – and not with water. We used foam. If it was a big barrel, you would direct the foam to the far side of the wall, and it would spread. As it spread, the fire would get smothered more and more. An electrical fire is different. No water, but no foam either, because electricity and moisture don't mix. So, you use dry powder, like flour from a pressurized container.

All these things you use somewhere in life along the line, but you learn it when you're nothing but a kid. Sometimes, it's amazing to people. They wonder, "How do you know how to do that?"

I had some good times, learned a lot, and experienced things that would serve me later in life.

~

Good Times

Once on the West coast, I went to a dance party. We'd docked near San Diego with the aircraft carrier, and we spent a few days there. Me and a friend – a guy I worked with on the ship – got roaming around. You couldn't go anywhere unless you took a cab or walked, but we got to know a couple of guys from there and one guy offered to take us touring around a bit one day. He said, "If you guys have time tonight, I'll take you up to a nice place."

I said, "What do you consider a nice place?"

He said, "Near the city of Hollywood," which is a part of LA, just northwest of the airport. "I'll take you up to the Hollywood Palladium. There's good music up there."

I looked at my buddy. "What do you think, Tyler?"

He said, "Yeah, I think so."

So, we decided to check it out. We wore our 'best white' uniforms - nice white, clean uniforms with 'Canada' written in gold letters on the shoulder pads. We stuck out pretty good.

When we got there, everybody was eyeballing these guys from Canada. We danced with a couple of the girls there, in the Hollywood Palladium, to the music of Les Brown and his orchestra.

I think more of that today than I did then. I was too dumb to know what I was getting into! I didn't really appreciate what was being presented to me. I was too young and off the farm, you know? But I think I'd like to do that again! It was a good time. I don't think the place exists anymore, and I don't think you can find a big time band anywhere, either. You'd probably have to go to Europe, or something. Anyway, it is just one memory that stands out during my time in the navy.

Another amazing memory is going through the Panama Canal to the Pacific Ocean. It took all day to go through the canal. Engineers on the ship had to take the gun sponsons off because they wouldn't go through the locks. They've restructured the locks since then because of the bigger ships that need to pass through.

But I really enjoyed that trip through there. Halfway

through, one side is open country, and the other side is like a mountain with all the jungle vegetation, like you'd see in a movie. Right in the middle of that, a huge waterfall came cascading down. It had to be four or five hundred feet high. It was like a hose just spraying the water out and then it became a stream running into the canal. It was really beautiful.

We see things that stick in our minds, over the years, that we enjoyed but we didn't appreciate at the time, the way we would today.

An English Adventure

I was one of six duty jeep drivers during my time in the navy. We spent six months in England in Portsmouth on radar refit, and I was out every day driving officers around. We'd take them out in the afternoon, and then around eleven or twelve o'clock at night the duty officer would bang on the door and say, "Captain so and so - or whoever - is in such and such a place and needs a ride home!"

Well, they certainly couldn't walk, so they had to have a ride, if you get my drift! They were all young guys, having their good times, just like all of us.

Of course, North American vehicles have the steering column on the left-hand side, but in England, you drive on the left-hand side of the road. So, the driver is up against the sidewalk. In the daytime, I'd drive nice and slow, and talk to people on the sidewalk, or wave on the way by, and so on. It was kind of fun.

While in England, we found a place for the average Joe to have some fun. It was a pub with live music and a dance floor. It had a restaurant, two bars and a balcony. If you had a partner, you could go up to the balcony and have a coffee or a sandwich or whatever and look down on the other people. They called it The Savoy.

I started going there, and I met this girl. We saw a lot of each other for the six months I was there, but I don't think we ever shared a kiss. We weren't intimate. There was no lust between us that you might expect or that you hear stories about. We were just good friends. She took me to different places to see the sights.

This was in 1953, and there were still lots of signs of damage from the war, so she used to take me to these places. I met her parents one day, and it kind of prevented me from being homesick, I guess. A lot of times we'd meet right there at The Savoy. Sometimes, we'd maybe have a beer, and enjoy dancing and that. If I couldn't take her home, I'd pay for the cab to take her home.

One night, we were out and we left the bar - me, her and another couple. There was a sea wall nearby, the top of which was a couple of feet wide. We were walking and talking, and all of a sudden, we slid right down into the water! I don't know if she knew how to swim or not, but she started screaming, and I grabbed her close to me and told her, "It's okay. Settle down. We're okay."

My buddy got down on his stomach; the girl he was with sat on his feet so he wouldn't slide down, and he reached down and grabbed my girl's arm.

I called to him, "When the wave comes in, you pull and I'll push." And up she went! Next, the two girls had to hold him down while he reached down for me. We managed to get out of there, but we looked like a couple of monsters that came out of the spaghetti swamp, with all the seaweed clinging to us and everything. It was hilarious. I asked her, "What's your mom going to think?"

She replied, "I know my mom, and if anything, she'll say how stupid we are. You deserve to get all wet!"

That could have turned out bad, but it was hilarious. Nobody got hurt. I sent her home in a cab. Now, I see God's hand of protection when a lot of people wouldn't. If you pay attention, there's not a day goes by when you don't get help from God in one way or another.

Chapter Four
GILLES OF ALL TRADES

ARMY RESERVES

I was in the navy for three and a half years, but decided I needed to get out. There were too many distractions in the navy; too many chances to get into trouble. I was young when I joined. My mom was alone on the farm, and I just wanted to get back home.

I got a lot of different jobs after leaving the navy; did a lot of different things, some of them quite interesting. Of course, I ended up being a long-haul truck driver for the majority of my life, but I had other jobs, too.

One thing I did was join the militia. I was in it for five years, as an army reserve. Of course, it wasn't my primary job, but I spent a fair bit of time with it, and learned a lot.

When I joined, they assumed, since I had been in the navy, that I must be a good drill sergeant. (I'd never actually made sergeant – I wasn't there long enough, and they don't use the term 'sergeant', anyway!) But they offered to make me a corporal if I did drill sergeant duties for awhile. After a few months, I was told to go into Bury, a small town only about 8 miles from the little town where I lived, and write the drill sergeant's course. They all knew I knew how to be a good drill master, so if I passed that course, then they'd give me another stripe, so I could actually be called 'Sergeant'. I thought that sounded interesting, so I went down, and I passed no problem.

From past experience at home, in the navy, and now this, I had seven different driving license qualifications for the department of defence, from a jeep to an army tank, and every different size of truck and vehicle in between. I was a tank driver instructor, a radio installer, and user instructor.

I gained some experience with radios and stuff, so I was in charge of that, too. Because they figured I had a good radio voice, I had to teach some guys how to speak into a microphone. It makes a lot of difference to people who are trying to listen!

When you get a bunch of vehicles out on a test, you use military communications equipment. People up on the dew line up north - NORAD - can pick these signals up. So, when you open up your net, whether you have two, six, a dozen, or whatever amount of nets opened up, you form that network and it has to have a name. Then, one of the first things you do is introduce yourself as military training, so that the military who are watching the borders will know what's going on. We don't want them to think our exercises are a real threat!

Another sergeant and I went to an army training base at Farnham, Quebec. We'd go up there for a month in the summer time for extensive training, since they had bigger tanks and different things for guys who wanted to get promoted.

We had a group of tanks that we could use, so we had to install our radios and get them working. That was always a good excuse for getting the tanks out and going through the bush! Of course, there were always officers who wanted to come along for the ride, so we got in good with the officers, too.

But you see, even though Farnham was a training camp for the Canadian army, they still had accidents. Accidents can happen anywhere. The one thing that we trained for there was how to cross a ditch with a tank. If the ditch was four or five or six feet wide, how would you cross that ditch?

The commander takes his position in the middle of the tank, and when he stands up, the top part of his body is exposed so that he can look around and see everything outside. Well, one day, they were doing this exercise, going across these ditches, and he directed the driver on how to approach this ditch. Unfortunately, he was way, way off. When the tank went through, one side fell into the ditch and it catapulted on its side. He got crushed between the tank cover and the ground.

It wasn't intentional, but these types of mistakes are things that we have to be wary about. Always pay attention. If the driver had been on the ball, things might have turned out differently. All the driver can see is a little slit. It's very narrow, but wide enough so that he can see wide. He could have said, "Sorry Sir, I can't take it at that angle." I mean, the commanding officer was as green as the rest of us. Afterward, they said, "Oh, we should have had a more experienced guy there." But hey, accidents happen. We don't like them, but they do. Ninety-nine times out of a hundred, it's in training, so you learn not to make these mistakes, but sometimes it's not pleasant.

One Mile Deep

I worked at a uranium mine at Elliott Lake back in 1958, after I left the navy. I was about twenty-two or twenty-three. We worked one mile straight down under the lake.

I enjoyed it. It was interesting, and there was lots to do. When you loaded up at one tunnel with a truck, you had to go to a different place - like a big room blasted out - and then down a tunnel to what they call the 'skip'. When one bucket was coming up, the other one was going down. The ore went to the surface to the plant to be processed. So, you drive in there, make your turn, dump your truck, and then go. Round and round. On one trip, I'd just gotten in there when right behind me - KERBOOM! A big chunk of loose rock fell right behind me, behind the truck. If I'd been two seconds later, I'd have been wearing it.

Again, when I think back, it was the hand of God that protected me!

The foreman asked if I'd consider working off the truck and start working on securing the ceiling. They needed a team of two guys going through the tunnels to prevent that happening again. You'd get a list of loose pieces that needed to be blasted, and remove them before they could fall on anybody.

I told him I'd been around some blasting a bit, but nothing like down there. But I was willing to learn.

He said, "Don't worry about it. You'll have a partner. He's a big Frenchman from Bellville, Ontario."

I think that old boy took me under his wing. He was twice my size and probably twice my age, too. But, we went around with a drill - a 'stoper", which is like a hydraulic leg run by air. You'd put a drill in the top of that and use the air pressure. Then you'd use a valve to get the drill turning with the pressure, and then we'd drill holes in this loose stuff and insert as much dynamite as we thought was necessary to break it up.

We got our regular hourly rate plus a bonus for every stick of dynamite we used without killing anybody! Stuff like that, if it's not handled right, can be quite dangerous. We worked on that for quite awhile, and it worked out really well.

After awhile, I started getting word from home saying, "You gotta get out of there. It's too dangerous. People are dying." So, I got talked into leaving, and took a job as a game warden, but I was at that mine from 1958 to 1960.

I suppose that part of the job was dangerous, but I enjoyed it. The uranium itself wasn't very dangerous. It would make a geiger counter go off - that's how they found it - but it needed processing before it was strong enough to do a person any harm. So, that wasn't the danger. It was more the blasting, and that made my family worried.

Some guys didn't like the idea of the water dripping, either, since you're under a lake! There were three mines under those lakes. The one I was in was called Denison Mines.

Interestingly, when I came out to Tumbler in the late '80s and working for Bull Moose, the guy that was in charge of the housing for the employees was the manager from the housing from Denison Mines from Elliott Lake! He remembered me, even though I didn't remember him. It worked out good, because he just gave me the keys, and I went around town looking at different places until I picked the house I wanted! But, that's a different story.

My Dad worked as a night watchmen while I was there, and then my brother also came there looking for work. We started down the shaft, and his eyes began to bulge the further down we went. He asked, "How far are we going?" and I when replied, "Oh, just a mile under the lake," he never said a word.

When the cage stopped and the gate opened, he got out, looked around and said, "This is not for me." He got back into the cage and went back to the surface! He worked for a couple of weeks on the surface, but then he left. It wasn't for him.

We often had different viewpoints on things. Anyway, all these things that happened are controlled by God. Either He's been protecting me or trying to teach me something. And you can't find anyone better qualified for teaching you than God!

Game Warden

When I quit the uranium mine, I became a game warden for a private club. A man who owned three farms rented one out so that these guys from the city could have a private place for hunting and so on. The club's leader was a colonel in the army. He's the one who talked me into joining the militia. He and my dad grew up together, basically, so it was almost like an old family reunion.

In those days, and in that area, the game warden patrolled the property. You had to answer to the provincial game warden for everything, but you were working privately. We patrolled the area that this club had in their control, and in order to hunt or fish elsewhere, they had to get permission from the farmers. They would go to the farmers and promise fire protection as well as make sure nobody messed with anything, as part of the deal to hunt on the property.

They made deals with a few farmers in the area, so it was good. In the winter time, I had a ski-doo and a shovel, and I went to these different cabins and had to shovel the roof off to make sure it didn't cave in. Things like that. In the summer time, you checked everyone who came on the property to make sure they had a hunting license. You'd check to see what kind of ammunition they were using, and give them a little 'spiel' about what they could and couldn't do and so on.

A lot of hunters could go anywhere, but if they didn't belong to the club, they weren't supposed to go on club land, since it was posted. There were a lot of things to look after, and it was interesting, but it wasn't enough full time wages to raise a family on.

~

HIGH WIRE ACT

I got into electronics when I joined the army. It led to working eight years with a guy who was an 'electronitian'. There's a difference between an 'electronitian' and an electrician. An electrician looks after house wiring and stuff, but an 'electronitian' was like computer work. Technical stuff. We'd call it something different today.

I was the guy that looked after all the antennas. There weren't satellite dishes in those days. The word wasn't even invented yet! Instead, people had antennas stuck on their roofs, and you had to put up towers and masts and what not. The masts were in ten foot sections, so if you had a 30 foot mast, you had one with two more ten foot sections inside of it. You'd take the one, put the guide wires on, then lean your ladder on it, and lift the other one up until it clicked in. Now, you're twenty feet high! Then you'd tie that one down, and so on, and before you raise the last one up, you'd put your antennas on. Then you had to orientate your antennas for the channels you wanted to receive.

The ones that had routers could be turned from inside the home. There were 'all wave' antennas that you could turn in the right direction to receive whatever channel, and there was all different types of wiring and that. I did that job for eight years, so I learned a fair bit. Later, we installed cable in a small town, so I got introduced to that, too.

Everything was run on tubes back then - both radios and TVs. When electronics started, in one TV set you had two 'half wave' rectifiers, which took the 110 voltage and converted it into 12 volt, or whatever the particular system worked on. Those rectifiers were pretty large! Later, they were replaced by two diodes. You could fit both diodes on the top of a dime, so it was a huge advancement. At the time, it was cutting edge.

I've always been interested in that type of thing. I did TV repair on the side, too. I carried a suitcase full of tubes for TVs and different gadgets for hooking things up - whatever the customer wanted. If I couldn't fix it there, I took it down to the shop and my partner would repair it because he knew more about it.

I did this at the same time that I was with the militia. The militia was only once or twice a month, not a full time job. I actually got a little business putting up antennas for the army, too.

Chapter Five
HARD KNOCK LIFE

GEORGIE

In the summer of 1965, my six year old son, Georgie, got hit by a car and was killed. It is definitely one of the hardest things I have ever experienced.

I was working fairly close to home, running a dozer back in the bush. Somebody came and told me I better hurry home because there had been an accident. So, I did, and there were still several people gathered.

He'd been hit from behind by a car doing between seventy-five and eighty miles an hour. That's way up there if you're looking at kilometres. His little body flew 174 feet through the air.

When I got home I said, "Where is he?" and they said the ambulance had already taken him to the morgue in the city.

I said, "I'm going down to see him," and everyone said, "You can't do that." I said, "Watch me."

My mother, my wife, and my pastor came with me. The pastor knew where the morgue was. When you pull into their yard, there's a big building with large doors. I pulled up in front. The morgue itself was right beside the house where the mortician lived, and he was up in the second floor, so he saw me pull in. I got out of the vehicle and was looking around when he came out onto his little deck and asked, "Can I help you, Sir?"

I gave him my name and said, "You've got my son in here."

He said that was right, and when I said I wanted to see him, he said, "I'm sorry you can't do that. He's not prepared."

I pointed. "He's in there?"

"Yes, sir."

"And I can't go and see him?"

"No, sir."

I said, "He's my son and you're telling me what to do? Yes, I can and I will. Just watch me."

I don't think any of us know how we're going to react in those kinds of situations. I got in my car, and I backed up to take a run at the building. He came down in a hurry, like a scared rabbit, and opened up the doors. He wasn't too happy about it, and told me I shouldn't be doing this, but he let me in anyway. I guess he could see that I meant business.

I went in, and it was really heartbreaking, but I had to see him, you know?

The young fellow who'd hit him had been coming home from work. He worked in a guitar shop, building guitars. We lived in between the town that he worked in, and the town he lived in, so he had to drive by our place every day, back and forth from work. He had not been drinking, just driving too fast and not paying attention, I suppose. He had just gotten engaged, and they were going to get married in the fall.

I could have charged him for manslaughter.

I had a lawyer - a guy I actually went to school with - and I told him about the case. He said he would come up and have a look at where it happened. He also had a six year old, so he came up the next day with his son, and he stood his son where my son had been hit, and backed up. Before he lost sight of his son, he was way back, on top of a hill. He said there was no way that guy should have hit my son if he'd been paying attention. Plus, from the speed estimate we got from the cops, he was travelling way over the speed limit. He asked me, "What do you want to do, Gilles?"

I had to think about it. The kid was planning on getting married. I needed a few days to think it over.

So, I went walking up and down the road. We lived out by a field, and I was walking and walking and my mind was going bananas.

My eldest son, Jimmy, had gone and picked up Georgie's little shoes off the side of the road where they'd flown off his feet. He put them in his school bag and walked around for two weeks with his brother's shoes in his school bag, carrying them on his back. I mean, things were pretty tough.

I finally made up my mind. It's not like I had never gone over the speed limit. You don't need alcohol in order to speed. Plus, the guy was getting married, soon. Why ruin his life completely? I'm sure he was suffering enough as it was.

I gave the lawyer the message that I wasn't going to charge him, and my lawyer wanted to know why. I said, "The only reason that I'm getting through this and staying sane, is that while Georgie's body was still in the air, he was already in the arms of Jesus." That was my steadying point. I'd think of that. "Georgie is safe." No more pain. He's with Jesus. That helped me through it.

It was the largest funeral our town had ever had.

The family of the one who hit him never came to see us or talk to us afterward, but they came to the funeral, and then they paid for it, too. They also sent a cheque, in care of my mom. In those days, for people living on the farm or working out in the bush, it was a fair amount of money, I guess, but of course, nothing could replace my son. At least an effort was made to show that they cared about what happened. It could have been one of theirs. It could have been anybody, and in the end it brought a lot of people together, actually.

∼

Another Son Lost

I lost two children. Losing children has got to be the worst thing in life for a parent. It's rough. Georgie was the first, but then my son Jimmy also died in 1990. It was another tragedy, as any death of a child is. He committed suicide.

I am convinced that drug abuse precipitated the suicide. He was married at one time, and had two little boys. His wife put up with him for a long time. At the time, he was up in the Yukon working. He got his Christmas holidays early, but when he got home, he found another guy in his home with his wife and two kids. He told me on his way back up to the Yukon that he didn't make a fuss or anything, because he knew he had a drug problem and knew she'd been putting up with it for a long time. He did get to spend that week with his two boys.

Then, he went back to work, and when he reached the camp where he was staying, he got into the drugs and alcohol even more. One night he called his buddy and said, "Get over here right now! I need you!" He told his friend he was going to end his life, but it was a plea for help, I think.

Later on, sometime in the middle of the night, his friend called me. He told me his name and told me what had happened. He said when he went over and opened the bathroom door, Jimmy was just sitting there. He said Jimmy had a gun in his right hand, and as soon as Jimmy saw his face, he reached around and shot himself in the left side of the head.

As soon as I heard that I thought, "He didn't actually want to die." He wanted to wound himself - get into the hospital, where his wife might come, and everything would be okay.

The medical team were getting ready to fly him down to Vancouver to St. Paul's hospital since he wasn't actually dead yet, and I got time off work to go to him. (I was working at a coal mine in Tumbler Ridge at the time.) I got to Vancouver and went down to ICU where he was laid out. He was all hooked up to life support, and as soon as I saw him, I knew he was gone. There was no life in his eyes. Everything was dead.

He had a small plaster on the side of his head. When I looked at that, I said to the doctor, "That doesn't look all that terrible."

The doctor said, "The gunshot's not what killed him." He told me Jimmy had enough drugs in him to kill two elephants.

What are you supposed to do when something like that happens? What are you supposed to say? Do you harden your heart?

To prevent a whole lot of talk and bad mouthing, we just said he'd had an accident at work, but the local people knew what had really happened.

Sometimes, when I think about all that's happened, I get teary eyed. Like, sometimes, when I hear gospel music, I'll feel like crying because the words they're singing, and the way they sing it, brings back memories of the past. Is that good for you? Is it bad for you? I don't know. I suppose it's a way of release to level out the stress. But no matter what anyone says, it still bothers you. I like to say I don't live in the past, but you live there more than you realize. I don't know what I would do if I had to do it all over again. If I knew what was going to happen, I probably wouldn't want to go there.

This pastor commented once, "After all you've gone through, you don't hate God." I said, "Why should I hate God? It's not His fault." I'm the one who's a sinner. I'm the one who's disobedient. I'm the one who deserves to be punished. I think God is just testing my faith to see if I have any or not.

∽

BROKEN HEARTS AND PROMISES

Yes, I've had my share of difficulties. The first part of my life was less about health issues, but I did lose six members of my family, and went through a divorce. I became a single Dad trying to raise all those kids, and still had to figure out how to make enough money to eat.

First, I'll list the family members I lost. Of course, there were my own two boys, but I also lost my nephew, my sister's oldest boy, who also got hit by a car only half a mile further off the road. The car that hit him swerved into the ditch where it hit him, and threw him onto my Aunt's front lawn. He landed face first, and there were elbow marks in the lawn. Again, somebody speeding and not paying attention, when all of a sudden, something appears and then 'boom-boom'.

Then, my Dad died in 1974 in a car accident. So, there's three car accidents and three subsequent deaths. In '86 my Mom passed away with Parkinson's and associated problems. In 2006, Betty's youngest boy passed away early one morning with asthma.

We have plans about how life is going to be; what we want to do, but it doesn't always turn out.

My first marriage was like that.

I was twenty when we got married. I had just gotten out of the navy, and, well, she got pregnant, so we got married. That was the right thing to do in those days. Her parents lived in Ontario. They were some of the most beautiful people I ever met. Good Christian people who would do anything for you, so it really hurt them - and me - when she left me.

We were married for 15 years. Then she decided all of a sudden that she didn't like family life anymore, and she took up with another guy who was divorced. He was a bartender, a pimp, a drug pusher, and just about anything else that you could think of. They've both since died horribly of cancer.

I suppose in some ways, she didn't have it easy. I was away working a lot, and she was home with all the kids. And I know it wasn't easy for her after Georgie died. She was on tranquilizers for a few days after he was killed.

Her leaving hit me really hard. I became a bum myself for about three years after she left. She left me with all the kids. Six kids, and a baby in diapers who wasn't actually mine. Jimmy was my oldest, then Paul, then Georgie would have been next. After him was Gilles William (my middle name is Arthur, so we called him Gilles William to separate the two of us), then my daughter Linda, then John, and then Mitchell who we called Mike. Finally, there was the baby, Dawn, who wasn't even my biological daughter, but I looked after her anyway.

As a single father, it wasn't easy. I tried hiring babysitters and housekeepers and what not, but most didn't last. Looking after that many kids was a full time job and it was hard to keep somebody. I had to quit my job and got some government assistance, but I didn't get half of what people get now-a-days and it wasn't enough, so I still had to try to work.

I made ump-teen trips from where we lived to Ontario, to

where my ex was with her boyfriend, to try and get her to come back. One time I even said, "You look after the kids and the house and everything is yours."

All she said was, "F you and the kids!" If she hadn't included the kids in her little "F you" rant, I might have kept trying, but that didn't sit well with me, so we figured it out on our own.

Eventually, the kids started growing up and leaving home, so the package was getting smaller all the time. We made out. It wasn't an easy thing to do, but it's just part of my life. I don't judge anybody anymore. There are a lot of songs that say hearts and promises are made to be broken. It's all part of living and learning, I guess.

Chapter Six
KEEP ON TRUCKIN'

BORN TO BE A TRUCKER

The first job where I actually got paid was driving a truck. I was only fifteen! I never even had a license!

Back in those days, they weren't as fussy as they are today. I drove a single axel truck, hauling five cords of soft pulpwood from the bush to the pulp and paper mill. I started riding with a guy and then started driving it home. Shortly after that, I was driving it alone on round trips. Things have changed a lot since then.

I had to drive in the mountains down in the States. We used to take our pulpwood into New Hampshire and Vermont where they had a paper mill, and you had to go through the mountains to get to the place. You had to make sure that your brakes were working and everything. It was a lot of responsibility for a youngster.

It seems I've been driving something my entire life. I drove truck off and on in between my many other jobs, and usually those jobs involved driving, too. The last job I had before I retired was driving a grader!

Driving is in my blood, I think. I like my own company, so I don't mind the solitary nature of long-haul trucking. When I started trucking, you never, ever drove by a vehicle that was stopped on the side of the road, because there might be someone who needed help. Now, things have changed. It's almost taking your life in your hands to stop, with all the rip-offs and danger. Yet, there are still a lot of good people out there, and lots more who suffer because of some people's selfishness.

During all my time on the road, I've never had any "chargeables". That means, I never got caught on drugs, alcohol, or had a serious accident.

Coming out of New Hampshire one time, there was a total white out. I came down the mountain from the mill. I was taking these paper rolls to Montreal where they'd make newspaper or whatever. I came down out of the hills, and there was a level area, but first you came to a T, and you had to make a left-hand turn. The way the snow was blowing and everything, it was pretty treacherous. Even today, I don't know how I ever made that corner! If I'd have gone straight through, I would have ended up in a real deep ditch.

I came around that corner, and the rig just seemed to go around all on its own. Ten minutes later, I was in this little town where I pulled over to the side, shut her down, had something to eat, and stayed the night. The next morning, the storm had gone by, and they'd cleared the road, so I was able to continue.

God has had His hand on the wheel right from the start. He's a better driver than I am.

Some Trucking Memories

When I finally picked myself up after my divorce, I wound up working for an outfit out of Edmonton, driving truck. It was 1977. This guy from Edmonton advertised for long-haul drivers, hauling cattle from Alberta to Toronto, and he gave a phone number. A couple of days later, I had someone drive me to the Montreal airport, and I was on the plane. I ended up doing other runs for him, and lived in Lloydminster, Alberta, working out of their terminal there.

I often drove to Toronto from Edmonton with lumber, pulling two trailers. I unloaded just north of Toronto, to a place they were building that was like a miniature Disneyland. They wanted me there at six o'clock in the morning. Well, guess what time that is on my Edmonton watch? Three o'clock!

At six o'clock their time, they'd be knocking on my truck door, and by the time my eyes were open, I'd have all the straps off, and the guy on the loader had emptied the trailer. Once I put my stuff away and got the papers signed, I'd be off to Hamilton where I'd go to the office and turn in the paperwork. Then I'd fuel up, and get the paperwork for a load of steel going back to Edmonton.

I used to make that round trip in five and a half days. I made better time alone than some guys did with two drivers. By the time I got loaded at the Hamilton steel mill, I would take a trip right straight back. Some guys would stay there the weekend. They'd rent rooms, get booze, meet some friendly girls... And they'd spend a thousand dollars! Then, at the end of the month, they'd wonder why my pay cheque was bigger than theirs. Well, I made two extra trips!

I had more fun doing what I was doing than what they

were doing. I had no hangovers. There were two things I didn't like - waking up with a hangover and an empty wallet!

I owned my own truck for a while, too, but sometimes it's good, and sometimes it's not. It depends what you're doing and where you're going. I'd sooner drive a company truck hauling oversized and overweight stuff, going into the oil patch and back in the bush and stuff like that. It's hard on equipment and it's expensive.

I bought a new Mac truck - a 2000 - at a Winnipeg truck show and they delivered it to me in Vancouver. It cost $155,000, so they aren't cheap. But, I was disappointed in it. It wasn't what it was supposed to be, so I got rid of it and kept driving a company truck after that.

I worked up in Fort McMurray, Alberta, for over a month one time. Syncrude was just being built, and we were hauling stuff off the railroad to take over to Syncrude. They called it 'Koke'. Another guy went with me and we alternated. One day you'd make three trips and one day two. It took us a month to transfer all that Koke. It all went absolutely perfect, so at the end of that time, the company took both our units, washed them up, and gave us a going away dinner to show their appreciation.

One guy from Syncrude asked me if I'd be interested in going out of the country. I said, "I don't know. Is it a holiday?"

He said, "Saudi Arabia. You'd spend a couple of years there, driving truck. You may or may not like it, because there are bandits and stuff, but they have security..."

I decided I was happy enough where I was, so that never materialized. Still, it was an interesting opportunity.

∽

Mechanics of a Close Call

I've had more than one close call, but somehow, God has managed to keep me safe all these years.

I used to haul oil up at Red Earth north of Slave Lake. Two big oil tankers behind. I made three trips per day - long days, but it was good. One time I came back after my first load for lunch, and the boss told me to watch out for a couple of guys who weren't qualified to be out there driving. They were in tandem truck, but the RCMP said they couldn't do anything because they hadn't actually broken any laws. They wanted to go up through the bush to where we was hauling out the oil. Not sure if it was a joy ride or what, but these two idiots were out on the road. Who knows if they were just clueless or on drugs or what, but anyway I was on the lookout. It was winter, not always the best in those conditions. Now, you can't just stop on a dime with a big rig like that and especially not on an ice road. I saw them going in, and then on the way back out, I came up over the hill, and there they were stopped right in the middle of the road. I couldn't slam on my brakes, but I did what I could to slow the rig down. I couldn't go to the right because there was a steep ditch, so I came right up to them and tapped their bumper. I pushed them back down the hill, and then I went left into the 'toolies' through some brush and what not. I came close to a tree, but missed it. There was no real damage except from the brush and that. I got on the radio and called for help, and they came right away.

My boss was really worried. He kept asking if I was okay and that; he knew I had a history of heart problems, so probably thought it might have been too much for me, but I was okay. Wasn't hurt at all.

We went into town, and then they sent two flat beds. They put my rig on one, and it went straight to Edmonton. They gave me a different truck for a day, but within two days I was back on the road in the original rig.

The RCMP said, "How did you know to do that?" He was amazed. It should have been a way worse accident. But I say, it was the Lord. It's just another case of Him looking out for me. He's my co-pilot all the way.

Once I was on my way down to Oklahoma City, working for a small trucking company. It was an older rig, but it was supposed to be in good shape. Unfortunately, one of the elements that controlled the fuel input and RPM of the engine crapped out.

I pulled into a rest area, opened the hood, checked it out, and there was nothing I could do. I got on the phone and called the guy who owned the truck, and he told me to go into Oklahoma City to buy a part. I thought I could do that, but it's a bit bigger than my backyard! I asked him where I was supposed to find the part, and who was going to pay for it? He told me to go to a Kenworth dealer, buy the part, and he'd pay me when I got back.

I was leery, because I don't think that was in the contract, but I figured if that was the only way to get home, I'd have to do it. I went into Oklahoma City, found a Kenworth dealer, and I told him about the problem. I said, "I hope it's not too expensive, 'cause I don't want to have to sell the truck to get the part!" He figured it wasn't too bad – I forget what it was now, but anyway, it wasn't that much money, either. However, he said it was going to be quite a bit for the labour. Forty bucks an hour, which was a lot in those days.

I said, "I think I have the tools I need, and there's only two things that need to be hooked up." I told him what I planned to do, and he agreed that was the right thing, so I bought the part and headed back onto the highway. It was a little bit tricky, because I couldn't regulate my RPMs, but I managed to get out to a rest area and opened the hood. Twenty minutes later, I had the new part installed, and everything was just fine.

You have to be careful of some of these companies that hire on a budget - a shoe string. They could take advantage of a person, but there aren't too many of them around anymore. He paid me, no problem, once I got home. I just see this as another example of how God was looking out for me.

∼

Vancouver to Miami and Back Again

My favourite route was probably a loop that took an entire month. Vancouver to Miami to Toronto to Calgary and back to Vancouver. Depending on where you picked up and dropped off, it was somewhere between 11,000 and 12,000 miles. That's more than 100,000 miles in ten months.

There's an interesting story about how I got driving this route. Ships came from the Pacific Rim - Asia, China and all those places - to Vancouver, loaded with dry goods destined for South America. The people that were handling all of this stuff could have taken it all the way by boat, but it would take too long. They'd have to go through the Panama Canal, or around the tip of South America in order to get from the Pacific to the Atlantic. So, you're looking at a pretty long trip. A plane would be too expensive, so the next thing is a truck. The people I worked for said, "You've been around a bit. What would you think about driving down to South America?"

I said, "Well... pretend you're back in the old west in the days of the cowboys. On the stagecoach they had a 'shotgun'." That's what they called the co-driver sitting by the driver. He'd sit there with a shotgun in his hands in case of bandits. We'd need that same kind of protection if we planned on going to South America with a truck. Criminals, outlaws - whatever you want to call them - would see a nice shiny rig coming along, and I doubt if you'd make even one trip and come back in one piece.

My boss asked, "Well, what do you recommend?"

I said, "Why not take it 'in transit' through the US? In Vancouver, load up a refrigerated trailer with the dry goods going to South America. Drive it to Florida, which is on the edge of the Atlantic, and drop the trailer at the warehouse door. You would have to go to the airport in Vancouver for customs' clearance for a transit load, and when you got to Miami, the same thing. At the warehouse, just give the guy there the paperwork he needs. You grab the paperwork you need clearance for, go to the airport in Miami, and get the 'in transit' paperwork needed. Once the warehouse is loaded up enough, they'd load a ship and send it down to South America from Miami.

It worked out great, and I loved it. I'm used to working alone, and I enjoyed my work.

That route had some interesting quirks. Vancouver to Miami takes a week. Then you have to sit for two days - one day to rest and one day to load. After that, it's only a couple of days to Toronto, but you end up sitting there for a couple or three days. Next, you stop in Calgary, unload and re-load, and then it's back to Vancouver. The whole thing takes a month.

Once in awhile for my trip back, I'd get a load of shrubs and plants. They'd pack that fifty-foot trailer full! There was no weight to the cargo - you'd never be overweight - but for the protection of the plants, and because they're taking them from a hot, humid area and going into a truck box, I had to have the reefer going so that the plants wouldn't have a bad reaction to the change in temperature. Most of the time I'd go right back to Vancouver with that load, so I'd unload in Miami, and the next day head right back to Vancouver. I didn't have to water them or anything, because they were wrapped in plastic if they needed moisture. But, when we got to Vancouver to unload, they didn't want me anywhere near those plants! They checked every plant to make sure there was no damage done, but I never had any problems.

I have a lot of good memories from my time as a long-haul trucker.

Chapter Seven
LONG-HAULS AND GOOD PEOPLE

LOVIN' THE LONG-HAULS

In my days trucking, we didn't use GPS. We just used maps, and most of the time, I knew where I was going. One time I was hauling equipment back into the bush for the drilling rigs in Alberta. You had to watch out when using some maps, because some people drew them backwards! East was west, and north was south! It was crazy. Once I had to back up for about a mile with a loaded trailer to get to the right turn off. It was after midnight, in the winter, and on an icy road.

I often used a Road Atlas for the long-hauls. It had all the states and provinces, and all major cities and so on. I have a whole book on LA - just LA - that's an inch thick! I got so I could get around better in LA than I could in Toronto, for one simple reason. In LA, it doesn't matter where you want to go, there's a freeway. You're always on the freeway. In Toronto, they still have a ways to go, although they're catching up. You still have to travel on small streets to get anywhere in Toronto. The same in Vancouver. I don't like driving in Vancouver! I shouldn't say that, but it's true! I guess I'm looking at it from a trucker's perspective.

At one point, I went to LA once a week. I'd take dry goods down and bring back groceries. That would only take the better part of a long weekend - four or five days. But my trip to Miami, Toronto, Calgary and back was always a month. Twelve thousand miles, every month - and I loved it because miles was money! To drive half a day and stop, and unload something and go another few miles... stop and reload... back and forth... That gets tedious.

Speaking of LA reminds me of passing by Guy Lombardo's estate. I used to listen to Guy Lombardo on New Year's Eve. He was broadcasting from a place near Riverside - not right in downtown LA, but east in Orange County. From Riverside, you take the interstate that goes right straight down and joins the highway that goes from LA along the ocean front to San Diego. About three quarters of the way down, there's a big estate on the left when you're going south - Lawrence Welk's place. A lot of famous people had their homes there. They had their own private golf course and the whole nine yards. That's where Guy Lombardo broadcast his New Year's Eve program from - the Lawrence Welk estate north of San Diego. It was the, "sweetest music this side of heaven". Man, oh man! I liked when he said, "Guy Lombardo and the Royal Canadians." It made me proud, I guess. I always tuned in as much as I could, depending on where I was in the truck.

I loved the long-hauls. I liked driving all the way across the country, from Vancouver to Halifax. You've got time to realize you're going somewhere before you have to turn around and come back! Depending on the route, you could run into four weather patterns - spring, summer, winter and fall - all in one trip! You had to be prepared, because you never knew what you'd come across.

I suppose it's why truckers get to know the best places to stop. You know the old cliche… if truckers stop there, the food must be good!

That's what I found about this one place. It's the largest truck stop in the world, and it's in the state of Iowa. It's called Iowa 80, because you take Highway 80. It started out as a tiny little place in 1964, but now they've increased their daily customer service to about 5000 per day! They have parking for 900 trucks - that's including trailers, 250 cars, and up to twenty buses. That's one big parking lot!

They've got everything – restaurant, a dentist, barber shop, chiropractor, workout rooms, laundry, movie theatre, TV lounge, private showers, several fast food outlets, plus rows of gas and diesel lanes. I've been to truck stops smaller than that, but it's a really nice place. It's like a family house. Lots of truckers need these services, and it's not always easy to get them on the road. They treat truckers well.

Anyway, once a year, this place in Iowa has a special 'do' for truckers to show appreciation. It's not a rodeo exactly, but it's a big gathering of truckers and their families. I'm not sure what the number would be now, but a couple of years ago it was on for about a week. They might get up to forty-thousand people, and for truckers everything was free! It's like, one good turn deserves another. Truckers support them, and so they put this on for truckers.

Tourists get good attention from them, too. It's a good place to go if you're ever in that area. It's probably 200 or 300 acres. I used to go through there a couple times a month for a few years.

Good People

We meet all kinds of people in this world. Sometimes we don't always get along, or see things the same way. But, when you look around and you really need it, it seems like there are always good people around.

On one of my trips down in the States, I was going through this little town. I can't remember the name of the place, because I stopped at so many places. But that time, I decided that before I hit the interstate, I'd have a good snooze so I could bring my log-book up to nice and legal, so to speak. It was very nice out – I didn't need to leave the rig running for air conditioning or anything, so I shut it down, and I was doing up my bookwork.

Right across the street there was a guy with a barbecue going. He looked at me a couple of times, and just as I was finishing up and I was going to lie down, he came walking over. He introduced himself and asked if I was planning to stay for awhile. When I said yes, he came back with, "How'd you like to rest on a barbecued steak?" Imagine! I'd never seen him before in my life!

Well, that's exactly what happened. I went across the street to his place, and he fed me a steak with all the trimmings. He offered me breakfast, too, if I was still there in the morning. I was long gone by then, but there it is - a total stranger - being so kind and doing something like that! It stays with you.

Another time, when my second wife Betty was with me on a trip, we ran into some more really good folks. Now, I admit that sometimes we form weird ideas in our minds about places and people we don't know. Let's just say it. We can be prejudice, even if we think we aren't.

Anyway, we went down south and delivered our cargo, and then we had to go to Birmingham, Alabama, to pick up a load of steel to deliver to Edmonton. Now, a lot of folks in Birmingham are black. Don't get me wrong. I've met a lot of really good black people, and I've also met a lot of not-so-good white people. But we were nervous, I admit it, probably because we didn't know the area.

It was somewhere between eleven p.m. and midnight, and I didn't want to park on the street because of the area. I saw that the gate was open, so I drove in, slow and easy. I got inside the gate, and I was going to shut it down, and go to sleep. My lights were shining across the field to this big warehouse, and there was a huge door there, that was open.

A man started coming toward us from the warehouse. He was so big! Half the size of my truck! He was huge, and his arms were the size of my legs.

He came up to the truck, and I rolled the window down, and he said, "Hi there! How are you?"

I said, "So far so good."

He asked if we were here to pick up a load. I said yes, and he asked where we were from, and I said Canada.

"Oh! You're from Canada? We've got a couple of loads for Canada. Which one do you want?"

I gave him my purchase order, and he looked at it and said, "Oh yeah, I got that right on the slab there. Do you think you could park this rig right inside the door if you had to back in?"

I said, "I'll give it a good try," and backed in no problem.

Once I was backed in, I was going to get out of the truck and take the chains and stuff off, but he said, "Oh, no, don't touch nothing. We'll do the work."

It was the middle of the night, but those guys loaded my truck; they had all the chains and boomers and everything tied down, and they had me check it out before they'd let me move, or anything. I asked if I needed to tarp it, and they said no. So, I said, "I guess I'll pull out of the way and wait until morning for my paperwork."

Immediately he said, "No you won't. Here comes the man that has your paperwork." All I had to do was sign a release, and we left Birmingham and pulled into a truck stop by about two o'clock in the morning.

I had been worried. It could have been a bad experience. But in fact, it turned out to be totally positive. God must enjoy helping out fools, because he's sure done a lot of good for me!

You can't judge a book by its cover, it's true. Sometimes we think we're better than somebody else, but we're not. Everybody is created equal. It doesn't matter what colour your skin is. It doesn't matter where you come from or what language you speak. He loves us all just the same.

∼

We're All Just Human

I've learned never to judge someone by the colour of their skin, their nationality, or even other people's opinions. God is the only judge, of course, but I like to form my own opinions about people, not just take someone's word.

Once I was heading down to Miami, and someone at dispatch told me to watch out for this certain neighbourhood once I got there. It was known as a Cuban and Puerto Rican district and he said, "They'll cut your throat and take your rig!" Well, I unloaded and then went to the place I was supposed to go with the next purchase order. There was a little bell over the door when I entered the office. Right away I noticed a sign on the wall with "Jesucristo" on it – the Spanish for "Jesus

Christ". A Hispanic looking man came out, and we did our business, and then I said to him, "I like your sign." He countered with, "Do you know him?" When I said, "Yes," it was like a family reunion! We were hugging and dancing around the room! He told me not to worry because the person who was going to load me up was a Christian, too. The guy who'd told me not to go there was all wrong. I got treated like royalty and they were the nicest people ever!

In New York one time I was driving around and couldn't find an access to where I was supposed to go. The streets were narrow and it seemed congested. To top it off, it was fairly late – about ten o'clock at night. Now, it was considered a "dangerous" neighbourhood at the time. I drove around the block a few times, but finally decided I needed to ask someone local. I spotted a bar and parked my rig across the street, then went in.

It felt like the whole place got quiet when I walked in. I looked around and quickly realized that I was the only white person there. Everyone was staring at me like I had come from Mars. I walked up to the bar and sat down and waited until the bartender came over. Honestly, I felt like David and he was Goliath. He was so big!

He asked me what I wanted, and I explained my situation. Then he asked, "Where are you from?" When I said Canada, all of a sudden, another voice with a real southern drawl piped up from somewhere in the back of the bar. "Did you say you're from Canada?"

This other man started walking toward me. I'm sure the ground shook! If I thought the bartender was big, this guy was massive! I mean, he was like seven feet tall and almost as wide! I don't think I've ever seen such a big man up close, before or since. The bartender looked like a kid in diapers compared to him! Anyway, I waited, not quite sure what was going to happen next. When he reached me, I stood up, and seriously, I was staring right at his belt buckle!

He said, "Mind if I sit by you, boy?" I wasn't going to argue, so he sat down. "Y'all are from Canada, you say?" I nodded and then explained my situation once again.

Well, I was surprised when he said to the bartender, "Bring this man a beer and bring me one, too!" He went on to tell me that his brother had moved to Canada a couple of years before and was always telling him he should move as well. He had so many good things to say about Canada, I couldn't have gotten in trouble if I tried! Apparently, all Canadians are the nicest people on earth!

We had a great visit for almost two hours. It turned out I had nothing to be afraid of, and in fact, I think he appreciated the fact that I wasn't scared to enter and ask for help, despite the fact that many white people wouldn't have done so. I'm not sure if the part about not being afraid is true, since I was, at first. Maybe I'm just too stupid not to care.

I don't think any of this was a coincidence, though, or even luck. I think God had His hand in it and might have been teaching us both a lesson.

∼

A Bee Story

One of the major things for truckers in those days was their CB radios. The truckers channel was always Channel 19, home channels were usually 15, and emergency channels were 9. CB radios are not used the way they used to be, but at the time, they could be the difference between life and death.

I was going down the highway somewhere in Alberta or

BC - I don't remember exactly where, but it was in the mountains - when this female voice came on the radio. (Some people liked to monitor the truckers' channel and talk once in a while.) Anyway, this voice came on and says, "Hey guys! Silence please! We have an emergency!" There was dead silence.

She said, "We have a trucker who just got stung by a bee, and he's allergic!" She gave his location, and asked if there was anybody out there with a bee sting kit that could get it to him on time.

Apparently, he had his window open with his elbow out, and he got stung. He stopped on the side of the road, but was going into shock.

Somebody said he had a kit and went there and administered the medication. It took a little while, but he recovered. Of course, everybody was so thankful.

There are people out there that we don't know - bad people and good people, but why does it take a lifesaving emergency to know that the good people are out there? I mean, that lady didn't have to do anything, but she did. She went out of her way to help, and it paid off.

Too bad we didn't see more of that. It does my heart good to think about it. God works in mysterious ways, His wonders to perform.

Chapter Eight
THINGS YOU COME ACROSS

THE GOOD SAMARITAN

Near Slave Lake, Alberta, there is a little town called Faucet. It is south towards Edmonton, near Westlock. One Saturday morning, I was coming back from up north with my truck and an empty tank trailer behind, when I came across a car up in the 'toolies' sitting up on a snowbank. I thought I better stop and see what was going on.

A lady got out of the car. She had an infant in her arms, and it looked like she needed help. When I got close to the car, I saw that the driver had his window open. Well, the smell would choke an elephant! He was very inebriated. There was the reason for hitting that snowbank. And her with a little baby!

The baby couldn't have been more than a month old, and it wasn't that warm out. I tried to talk to the driver, but he wasn't making much sense, so I talked to her instead. She was shaking from the experience, so I asked her if she'd like to get into the cab of my truck with her baby, to get warm. It was a 'cab over' so behind the seats there's a bunk. I helped her across the snow, and she put the baby in the bunk while I tried to figure out a way to get them un-stuck.

I couldn't pull it out with one chain, so went back and I told the guy what I was going to do. "I'm going to pull you back with one chain, and then I'm going to take off the first chain, and hook up to the second chain." I told him, "All you gotta do is leave it in neutral, and hang onto the steering wheel so that you don't turn sideways and get turning the wrong way."

In his drunken state, he said, "Okay!"

So, we got that worked out, and I got the second chain on, and we got them back onto the road. The woman had gotten out of my cab with the baby, and was nearing their car. I leaned down to grab the chain from my bumper, and then I was going to take it off his vehicle. As I reached down to grab that chain, the guy tried to take off with his car! Needless to say, he never broke the chain, but he certainly made a mess of my left thumb! My thumb had been between the chain and the bumper, and it was really messed up - bleeding like crazy. The poor lady started going berserk.

I had to tell her to calm down. Then, I reached into my truck and got a bunch of paper towels wrapped around my hand and said, "Everything will be okay." I managed to get the chains put away and everything, but then she insisted, "I'm going to ride with you to make sure you don't pass out from shock."

She had the right idea, but maybe she was using it as an excuse not to have to drive with him. He was still drunk. So, we drove down the road, and she kept staring at me, keeping watch. I was smoking at the time, and she made sure I had a cigarette in one hand while also trying to drive with the same hand. The other one was hanging down, bleeding on the floor!

We made it to Faucet, and right beside the road there was a little restaurant. I had stopped in there a few times and knew the waitresses and staff, so she rushed right in there ahead of me and explained what had happened. (Her man stayed in the car. He never moved!)

When I walked in, the waitress took one look and said, "All I've got are aspirin. Do you want a handful of aspirin, Gilles?" I said, two or three would do. I told them I'd drive into Westlock, so, the waitress phoned the hospital so they knew we were coming, and that lady rode with me in the truck all the way to the hospital. Her husband followed. When we got there, she banged on her husband's door and told him he had to come inside - now!

He wasn't in much shape, but anyway, he did as he was told. When we got inside, the doctor took a look and said there wasn't much he could do with it but clean it up. He said he could sew it up, but tomorrow morning I'd need to go to the hospital in Edmonton where they'd re-break it so that it could be straightened out and be of some type of use.

Anyway, I left with my truck and trailer, and they left. I don't know where they went, but I thanked her for staying with me through it all.

When I got to Edmonton, I parked the truck. I was staying in a hotel on 18th Avenue. Early Sunday morning there was a knock on my door. I'd forgotten that on Sunday mornings a bunch of local guys played hockey together. My buddy came in and asked if I was ready to play hockey.

I showed him my thumb, and he just about passed out. He said, "I'm going with you! I'm taking you to the hospital." I told him he didn't have to, but he insisted. We got to the hospital and they admitted me.

On Monday morning at 8 o'clock, our safety supervisor

came into my room with a huge basket of stuff. He said, "You did something that not a lot of people would do. We want to show our appreciation on behalf of the company. This will get out there. People will say, 'A guy from Cascade Carriers looked after them'." I guess it's true that word gets around, and it was good for the company name. They really appreciated it, and they really looked after me.

I was off for about six months with that thumb. I got workman's comp, so that worked out alright, and when I was ready to go back to work they gave me a brand new truck and trailer!

∽

Come What May

While I was on workman's compensation, I wasn't really able to drive truck, but I did hang around with some of the other guys I knew. A friend was going up to Fort Nelson, BC, with an old company truck loaded with cement bags and equipment. He asked me if I wanted to go along. I like to keep busy, so it seemed like a good idea.

We stopped in Dawson Creek for something to eat. Maybe it was day-old soup, or something, but shortly afterwards I started to feel sick. By the time we got to Fort St. John, less than an hour away, I felt really ill, but we kept on going. When we got to Pink Mountain, we stopped at the truck stop there. I wasn't hungry, so didn't order anything but a glass of milk. However, my buddy asked if I'd like a bite of his sandwich. I took one bite of that sandwich and had to run out the door to puke!

He finally realized that I was genuinely sick. I guess it was food poisoning. Vomiting made me feel quite a bit better, so I offered to drive for a bit so my friend could have a little snooze. I ended up driving the rest of the way to Fort Nelson. At that time, there was no bridge on the outskirts of Fort Nelson like there is now, so I had to drive across the river on an ice bridge. From there, I wound all the way out to the camp, in the bush. I drove to the camp, unloaded the trailer, had a snack, got back into the truck, and started home. All that time, my friend was still sleeping! Halfway home he woke up and couldn't believe that he'd missed it all.

It was just the way I liked to operate. If a job needed to be done, I did it. Not too much could stop me.

On weekends, when I lived in Lloydminster, lots of the guys would get together at the bar after hours. Sometimes it got rowdy, especially on weekends, because work crews came into town off the oil rigs. They'd get drinking and get mouthy, and that.

One time, I was sitting there talking to my buddy, when a guy came at me with a pool cue. I'm not a big man, but I don't believe in taking guff for no reason! I got up and hit him. Unfortunately, he had a lot of friends, and they beat me up pretty bad. At one point, a guy was stomping on my face with his boot.

The cops came, and my friends got me away from there and cleaned me up as best they could. One of them asked me where I wanted to go, and I said, "I'm taking a load to Saskatoon."

They thought I was nuts, but I went right then and there to Saskatoon! It was on my schedule, and I wasn't one to shirk my duties or make excuses.

The next morning, I was a bit late getting back to our terminal. My boss had heard about the incident the night before and was getting worried. It wasn't like me to be late.

When I got to the terminal, he forced me to go to the doctor. Turns out, I had a concussion! I'd also lost most of my teeth from the boot applied to my face, which is why I now have dentures.

It could have turned out a lot worse. I could have died, I suppose, but to me it is just another example of God's plan and His way of looking out for me. In this case, I learned a valuable lesson: If you're going to act stupid, you're going to look stupid! Lesson learned.

∼

An Unpleasant Christmas Eve

Being on the road so much, I've seen my share of accidents. Bodies with bones sticking out and blood... When I stop at an accident and see all the damage and hurt and death, I think, "That could be me." Even now I pray for safety whenever I drive.

My boss's father came with me on a trip one time. It wasn't far. I was going to pick up a load of lime. It was Christmas Eve, around suppertime, and at that time of year it's pretty well dark by then. We were coming around a corner and here was this car crash. One car was going this way and the other coming from the other way, and they met on a curve, head on. There was nobody else there, yet. It had just happened.

I parked crossways along the road to prevent others from crashing into them. My boss's father didn't want to get too close to it, so I told him to stay and watch for traffic.

Oh boy... What a sight. A six year old girl was squeezed between two seats in a standing position, smothered because she was squeezed to death. Her mother was in and out of consciousness. It was bad. Alcohol and beer bottles were all over the place.

By this time, someone else had come along, so I had him help me carry the mother out of the vehicle. She had broken legs and arms. Somebody had a blanket, so we used a coat or something for a pillow, and laid her down. She had a baby - a little infant - in her arms, and that's what saved her. The baby took the shock. She kept screaming, "My baby! My baby!"

The other guy with me said, "Don't worry about your baby. She's dead."

Right away I said, "Whoa, whoa! You don't talk to someone like that in her condition."

He didn't appreciate me talking to him like that, so he left, but by then, others had come. Finally, the police came, so I said to Fred, my boss's dad, "Let's go." I'd given my statement and it was okay to leave.

These are things that stick with you. You think about them. It could very well have been me in that other car - the one dead or all broken up. And yet, through all the years, and the miles, and vehicles, God has protected me.

∼

Being Bilingual

Being from Quebec, I am fluent in French. I haven't spoken much French since I left Quebec, when I left in the summer of 1977 and went to Edmonton for work. However, being bilingual has its advantages.

Even though most folks speak English in the west, there are little pockets of French spoken in unlikely places. If you go into the middle of Alberta near St. Paul, a lot of those farmers speak French and they're not from Quebec. When I first came out west, I was driving a truck for a company out of Edmonton. This one Saturday, the dispatcher asked if I'd like to work on a Sunday, since most guys didn't want to. I was okay with it, so he loaded a flatbed trailer with bags of cement. I was delivering it to St. Paul, and he drew me a map of how to get there. He told me to be there early in the morning so they could unload it.

I got out there early, as directed, and found the place, no problem. I pulled into the yard and there were two or three of them there. They started waving their arms to come toward them, over by a shed. I did, and when I was going to get out and undo the straps because it was tarped, they said, "No, you don't have to do that. We'll untarp and unload. You just rest up."

As they were working, they started speaking French to one other. Of course, me and my big mouth, couldn't be quiet. They made some remark, and I came back to them in French. They all looked at me, and everybody stopped dead.

"How come you know French? Where do you come from?"

So, I told them. I had just come from Quebec.

I kid you not, that was like dropping a bomb! Those were the days of the separatist movement in Quebec. The FLQ crisis and all that. With all the separatists, and all the turmoil happening in Quebec, they wanted to find out whether I was a separatist or not. "Tell us who you are!" they said - and not in the friendliest tone!

I convinced them that I wasn't a separatist, and all of a sudden we were friends again! They didn't care that I could speak French and knew what they had been saying, they just didn't want to have anything to do with me if I was a separatist. In those days, Quebec was the only place in Canada that you heard about separatism, and they weren't in favour.

I had family back in Quebec, of course, but I didn't have any family who were pro-separatist. The biggest part of that movement was around Quebec City and Montreal – in the big cities. It was basically a whole bunch of young people looking for something to do, as far as I could tell, and then it spread.

I know that now there are people in other parts of the country talking about separating, but I hope it doesn't come about. I don't think it would solve anything. There is never a quick fix.

I don't like it when I hear people say, "I don't like that person because they're not from here," or something like that. I say, "Neither are you. Most of us are from somewhere else." Some of my ancestors came from Scotland, but others only spoke French and couldn't speak English at all. We shouldn't base our views on whether someone comes from a certain place, or only speaks a certain way.

It reminds me of another story. I knew a doctor's daughter from another small town in Quebec. They only spoke French. She and her girlfriend decided they were going to go on a trip across Canada one summer, from Quebec out to the west coast and back. Her father, the doctor, said, "You realize that when you get out of this place, you're going to have to start speaking English."

She said, "Never! I hate English!"

Well, they were gone for a couple of months, and when they came back, she'd changed her tune. She learned to speak English, and she enjoyed her trip. In fact, she said she hated to come back! To me that's a message that we should all listen to. Our pre-conceptions are not always correct!

I knew another fellow like that. It happened near a place called Thetford Mines, Quebec. That's where the asbestos mines were at Black Lake. (My dad used to work there, working with dynamite. He was pretty handy with explosives.) Anyway, a guy I grew up with and his brothers started building travel trailers there. They built tent trailers. I was driving truck for them, delivering the trailers.

Once in a while I'd have to take a trip to Newfoundland. I'd drive from Thetford down to the coast, and then take the ferry across to Newfoundland. Once you got across, you do like a horseshoe. You'd go way up island, and then turn and come on back down to St. John's. It was routine.

We had one guy driving truck who couldn't speak English. He was also scared to death of water. He only made one trip. He slept on the cement bench in the load up area, and then, because he couldn't sleep going across in the boat, he found another bench and had another snooze. He couldn't speak to anybody, because he couldn't speak English. He could follow a map, so he knew where he was going, but he only made that one trip. After that he said, "You guys can have that!"

I said to him, "You know, there's another world out there besides Quebec."

I've been lucky, I guess, that I could speak more than one language. It's come in handy in my life.

Chapter Nine
SECOND CHANCE AT LOVE

BETTY

I met Betty in Edmonton in 1980. She worked at the front desk at a hotel that was close to our trucking yard. They'd allow trucks - no trailers - to park in the hotel parking lot in the back. They had rooms that were cheap, a beautiful big dining room, a coffee shop, and three bars, and our company got a cut rate on rooms. Anyway, she and I eventually got together, and we got married in 1982. At first, we lived in Edmonton, but we moved a bit after that. People said it would never last. They gave us six or eight months, tops, but it's been forty years. That's a pretty long six months!

We spent our honeymoon down east. I took her to meet my family, and she has a brother in Vermont, so we went there, too. She was able to meet my mom before Mom passed. We came back up into Canada at Niagara Falls. My daughter had rented us a room there, so we had a wonderful time. It was good.

She had three children from a previous marriage: Dean, Kim, and Kenny, who died of an asthma attack in 2006 when he was 36 years old. So, between us, we had lots of family, but we never had any children together.

I miss her more now than I ever did before. She got dementia in her later years and finally, I couldn't look after her properly, anymore. It was a very difficult decision for me.

All the years I was trucking, I phoned her around six o'clock every night. We were used to being apart. We lived alone a lot of the time, but we were never really alone. But now, after I took her to the nursing home at Campbell River and left her there, I learned that you don't have to be a baby to cry. It hit me pretty hard. I know she's getting good care and she's content. I won't say she's happy, but she's content. But, still, it's a very hard thing for me after so many years.

I always appreciated what Betty did while I was away working. The house was in good shape, there was food on the table, the bills were all paid, and she was in a good frame of mind. Everything was good. And I appreciated all that, but now that she's no longer here, and it's all up to me, I'm beginning to appreciate her even more, because I have to do all the things she used to do, and I didn't realize it! Women deserve way more credit! I don't know if that's sentimental or what, but I really mean it. We just don't take notice and we should.

God's Healing Hand

Betty has been good to me. She was always thinking of someone else instead of herself. It was her, "Yes," that got us both back into going to church. What with my nasty divorce and people dying and so on, I'd been backsliding in my faith; going downhill.

One day I said to Betty, "What do you say we go to church

today?" and she said, "Yes." Just like that! I thought, "What?" I wasn't expecting that! So, we starting going to church.

In 1990, she slipped on some ice coming out of a store and she broke her hip. She was taken to Dawson Creek to the hospital (we lived in Tumbler Ridge at the time), and some old horse doctor got a hold of her and put a bunch of pins into her hip. He said, "That'll fix you up for life!"

Well, eleven months later, she was back in the hospital getting them all taken out because they'd all gotten loose. It was never really fixed properly, so her hip healed up a bit shorter on one side. She had to wear an elevated shoe after that.

Anyway, she smoked at the time, but she quit smoking while she was in the hospital.

Then in 1992, she was in the kitchen one Sunday, and I was in the dining room watching TV. Oral Roberts was on. I used to like to listen to him. At the end of the program, he always said a prayer for the people watching at home. To make it real, he told people watching at home to reach out their hand toward the television.

I'd been sitting there listening to him, enjoying a cigarette with a full ashtray beside me, and I thought, "Well, I'm going to try it." I'd already tried to quit a couple of times, but I kept going back to it.

So, I stretched my hand out to the TV and prayed with him. Then I got up, took the tobacco and the ashtray and the whole thing, and put it in the garbage. Betty said, "Is that it?"

I said, "I hope so!" That was in '92, and I haven't had a cigarette since. To me, I didn't just quit. I was delivered. There's a difference.

It's like another experience I had when I got the news that I had prostate cancer.

I came home and told Betty. We sat on the living room couch, and we talked a bit, and we prayed together, and we laughed together, and we cried together. When she asked me why I was crying, I said, "It's not for me. I have kids who don't know the Lord."

Afterward, she went to finish making supper. I was still just sitting there and all of a sudden - I still don't know how to explain it... I call it a freshness that started at the top of my head, and went through my system, and went all the way down through my entire body. It wasn't just a quick thing. It lasted a good eight or ten seconds. That freshness went through and then, whoosh - out the bottom! And something told me I was free of cancer.

The plans were already made for treatment and that, so I went through with it, but I believe I was healed that day.

On the Road With Betty

Once in a while Betty would come with me on my trips. One particular time, I forget exactly all the places we'd been, but we were headed down through Texas - probably going to Houston, and we got talking. "How do you like this place or that place..." and Betty said, "From all the places I've seen, I'd have to be forced to live in the USA. Every place seems so dirty! The streets are dirty and there's garbage everywhere."

I said, "Well, if you were forced, where would you accept?"

She thought for a minute and then said, "San Antonio, Texas." When I asked why, she said from what she'd seen, it was clean. There's a truck route that detours around the city and one that goes straight through, so I took her both ways. She enjoyed the one that went straight through the centre of town. Everything was clean. I think it's interesting that now, after all this time, San Antonio is the headquarters for Cornerstone Church, John Hagee's ministries. We always enjoyed listening to his ministry together. It entered my mind that this wasn't a coincidence. It's one of the good memories of a trip together.

On one other trip, she took a map of Canada and the US and drew lines to all the places we'd been so she could say we'd been there. She wasn't much of a traveller, though. She was a home body. But, she enjoyed the trips we did go on together, because I always took her somewhere new. I took her across the Hoover Dam at night when it's all lit up, and boy, was she impressed with that. When you get to the other side, there's a flat area for tourists where you can stop and take pictures and what not, so we stopped there and got out and saw all the lights.

A lot of these places you enjoy when you visit, but you wouldn't want to live there. She wasn't impressed with the Grande Canyon, for some reason. I always liked Flagstaff, Arizona, because I read a book about an old trapper or somebody who lived in that area. When you come into Flagstaff itself, it's on the top of a mountain, and you have to go down into the valley to get to the high desert heading toward Phoenix. It can be winter in Flagstaff and summer in Phoenix! After you go across that high desert, you drop down again to the lower desert, and that's the level that Phoenix is on.

There are these Joshua trees there. We went through the Joshua forest, and oh! She had to stop to take pictures, and she wanted to take one of their needles as a souvenir. But when she touched it she said, "Ow! That thing bites!" They're sharp, and they're stiff, and she put a little dent in one of her fingers!

But those are the types of things she enjoyed, because she

likes flowers and plants and things. She loved going through Florida, because of all of the birds and that! She'd say, "Oh, I saw one of those in a movie!" And there was one right there!

One time we went through Alligator Alley in Florida in the Everglades. When you're coming from the west, I think it's Highway 25 that goes on the west side of Florida all the way to the Everglades. Then they've got a narrow road going right into Miami. There are different places where you can stop and 'rubber-neck' with the alligators and what not. On the east side is Highway 75, but you can still get into Miami. Anyway, she really enjoyed that, but she wanted to make friends with this pair of eyes she saw in the ditch where we stopped. She got her camera out, and she was going to go over, and the eyes started coming toward her! She made a quick U-turn, I'll tell you! It was definitely an alligator. There's a ditch only on the one side, so all the 'gators are on that side, in the ditch.

People have airboats there that you can rent a ride on - different things for tourists as tourist attractions. I always found Florida interesting, and I went through there many times.

Another time, when she heard we were outside Nashville, she perked right up. We went into this place for something to eat, and of course, there were all kinds of posters about famous people, country music stars, and the like. They also had a little gift shop, so she wanted to look around. I stayed in the truck so that she wouldn't feel pressured. If she saw something she wanted to buy, she could just go ahead and buy it. She came out with this little lapel pin with "Jesus" on it and gave it to me.

It was a special gift. I forget which pastor it was who said it, but he said, "With that pin, you never have to open your mouth to talk about Jesus!" I guess that's true.

Chapter Ten
FOR BETTER OR WORSE

MOVING ON

In our forty years together, Betty and I moved to several different places. We started out in Edmonton, but then we moved to Tumbler Ridge, BC, for four and a half years.

The first time I visited, I went to Tumbler alone, because Betty was still working at the hotel. Friends of ours lived there - the daughter of one of Betty's best friends in Edmonton. I'd been driving truck up in Red Earth Creek, a hundred miles north of Slave Lake in the oil patch. I worked two weeks in and one week out. On a typical day, I ate breakfast at five AM, made one trip, and was back for lunch. I'd make another trip and then be back for supper. Then, I'd make another trip after supper, and get in somewhere around 11 o'clock or midnight. That was my day - every day - for my two weeks while at work. So, I put in a lot of hours, and I made good money. I didn't mind the work, either, but these friends of ours from Tumbler said, "Gilles, why don't you come to Tumbler and work at the mine? You'd be home every night. You'd work seven days, and then you'd get four days off."

It sounded good, and it made sense. Our friends offered to talk to the person hiring at the mine. Later, I got a phone call. I'd already warned my employer in the 'patch' that I might be getting a call for a different job, so not to feel bad if I gave my notice. When I got the call from Tumbler Ridge, I decided to take a run down and see the place first.

The guy who was running Bullmoose Mine at the time had been told about me - that I was a good, experienced driver and such. When I went in to see him, he seemed surprised. I'm not a tall person, and he said, "Oh! From what I heard about you, I thought you'd be about ten feet tall!" They hired me on the spot, and I worked there for four and a half years.

I did a good job, but when you're used to working alone, it's a different world, right? So, Betty and I talked about it, and I ended up getting a trucking job out of Langley, BC. I gave my notice, and we left Tumbler Ridge one January morning. It was about 45 below zero and it was dark as night! I drove the U-Haul, and I couldn't shut it off, because I knew I'd never get it started again if I did. Betty's son Kenny was up helping us, so he drove my pickup down. Betty rode with him most of the way, but she also rode with me some, too. We made it to Langley, and I was given a truck to drive right away.

Now, I was thinking I might like to own a truck of my own. I said to the owner of the company, "It seems to me when we talked that you said I could buy one of your trucks if I had a mind to."

He seemed to be in favour, so we agreed that I'd take a few trips and see how it went, first. If I liked it, I'd buy one of their trucks, which is what happened.

Then we bought a place in Mission, BC. We had rented for a year, but this one time, when I got home from a trip, Betty came to pick me up and drive me home. She turned a different way than was normal, and I said, "Where you going?"

She said, "Oh, we're going home."

I pointed. "But we live that way."

She said, "Not anymore!" She'd bought a trailer in this trailer park while I was away. She was independent like that, and it was a nice enough place.

We moved back to Tumbler Ridge in 2000, and have lived there ever since. When we moved to Tumbler the second time, I ran the grader for the Department of Highways. I did that for eight years, and I enjoyed it almost as much as I did trucking.

We'd come on holidays to see my daughter and my grandson, who were both living in Tumbler Ridge at the time. It was during that time, around 1999 and 2000, that they were practically giving houses away in Tumbler. Betty and my daughter Linda went carousing around looking at houses, and they saw one that Betty liked. She wanted to move back, so we gave a $1000 deposit and had a month to make up our minds. We did a fair bit of work to the place, and I built a garage.

Betty was always busy and puttering. We had a good run; a good life. To start out, like everybody else, sometimes we didn't agree on everything, but that doesn't mean you throw in the towel. You discuss it; find an alternative. Give a little and meet in the middle somewhere, and it always worked out.

You might be wondering why we moved so much and why I didn't stick with trucking if I loved it so much. Well, to be honest, by the mid '90s, I had trouble keeping work because I started having heart attacks. I've had five heart attacks, but that's a story for the next section.

It's enough to say now, that God was directing us wherever

we went. I know that my faith keeps me going, because it tells me that God is looking after me, and will continue looking after me. After all, this world is not my home, I'm just a passing through!

A Matter of the Heart

My first heart attack occurred in 1994. Previously, on October 29, 1993, I had kidney bypass surgery, and then eleven months later, I had my first heart attack. It happened at the Canada-US border crossing at the Peace Arch. The company had to send somebody there to pick up the unit and load, which they didn't like very much.

A year after that, I had my second heart attack. It happened at a steel plant in Edmonton while I was on a run there. We were living in Mission, so Betty drove to Edmonton to be with me. She had an aunt in Edmonton, so she stayed with her aunt and came up to the hospital every day. I spent ten days in the hospital.

The company I worked for said, "You've had two heart attacks already. We have to let you go." I guess they figured it was too risky.

It was hard getting and keeping a job after that. I never knew how long it was going to last, because it seemed that I kept having another heart attack every one to three years. I've only had five of them!

After one heart attack, our doctor in Mission said, "We've got to admit you to the hospital here. The only way to get you to Vancouver, is to have you in the hospital here." So, I was in Mission hospital for two days, and then they sent me to Vancouver and that's where I had my open heart surgery. It was supposed to be for August first, but the lady doctor that operated on me was a heart transplant specialist, and the night before she was supposed to operate on me, she got called out for an emergency transplant. She was in no shape to open me up in the morning at 6 AM! She came and apologized, and told me that it was my turn the next day, no matter what.

I had three or four bypasses. I've also had stents put in. I've been opened up like a cow being butchered - down each leg and behind the knees from my kidney surgery, and then from my sternum up to my chin for the heart surgery. I've got a zipper from my chin to behind both knees! People wonder and just shake their heads. They say, "You've been through all that and you still wanted to go to work?" Well, when you're not too lazy, it's good to get up and do something!

I knew a guy who had a bunch of refrigerated units that he ran down to California, mostly to the LA area. I told him if he wanted to hire me, I'd drive one of them for him. He asked me, "When can you leave?" and I replied, "When do you want me to leave?" He said, "Yesterday!" So, I worked for him for quite awhile, and things were going pretty well, but I always had it in the back of my mind... When's my next heart attack?

Five heart attacks? Lots of people shake their heads! They think I don't know how to count! Nobody survives five heart attacks! Plus, I had cancer... plus a kidney by-pass? No way, they say!

Well, it's very simple if anyone wants to find out the truth. Just call up the doctors or look at the medical records. During a three year period, I was in eight different hospitals between Alberta and BC. Call the hospitals! They've got records. It can be verified. Even if it's hard to believe, it's 100% true! Trust me, I know, because I went through it.

It's just another example of how God and His angels have

been watching over me my whole life. I guess He had plans for me, or I would have been dead already. As it is, I'm still around. Maybe He's not through with me yet!

Chapter Eleven
FINAL THOUGHTS

ONLY ONE THING...

I've lived a full life. I count it a privilege every day that I'm still tickin'. It's another day to tell someone about Jesus; to maybe have an impact on someone's life. You just never know what you might say or do that could affect someone.

Not long ago, I had a chance to minister to someone on the phone. It was a lady representing one of the political parties. I shared the gospel with her, and she was crying when we hung up. She told me no one had ever explained their faith to her like that before. Praise the Lord! He does have a plan and purpose for our lives. Really, we are here for only one a reason - to glorify Jesus.

I'm not perfect, but I know the One who is. That's what it all boils down to.

Gilles Goddard - 2020

A VALENTINE FOR GILLES

The Newspaper Article That Started It All

Gilles, an 85 year old friend with a history of heart troubles, was in the emergency ward last fall. Nothing new for Gilles. He's had multiple heart attacks, but still keeps on ticking like the soldier he is. In typical Gilles fashion, he was chatting up the nurses as they attended to him.

One of the nurses was new to the community, so he asked if she had been to see any of the spectacular scenery in the area. Tumbler Ridge is famous for its waterfalls and other scenic spots, one of only a few Global Unesco Geoparks in North America. She said that she hadn't had time yet, so, again, in typical Gilles fashion, he offered to take her, sometime.

I think he was surprised when she took him up on his offer, and so they met at the appointed time and place. She brought along two friends, also new to town, and apparently, they had a fantastic time. Eighty-five year old Gilles and his three young female companions toured some of the closer scenic spots accessible by car, and a good time was had by all.

A few days later, the 'girls' brought Gilles an envelope. In it was a laminated selfie of the four of them taken on their adventure. He was touched. "But there's more," one nurse said. Surprised, Gilles found a lovely card, thanking him for his generosity and time. Again, he was touched. "There's still more..." she insisted.

Puzzled, Gilles looked in the envelope again. To his surprise was a fully paid airplane ticket worth more than $750!

But the most touching part of this story is yet to come. Recently, Gilles was forced to place his wife of forty years in a care home on Vancouver Island. Despite his best efforts, her progressing dementia was too much for him to handle by himself. Nursing homes with good care are hard to get into, and furthermore, they couldn't find anything close to home. Gilles made the decision to put his dear wife in a home 2000 kilometres away so that she would be nearer to her children. This meant that he only saw her infrequently when he made the twenty hour trip by car every couple of months. Until he could sell their home in Tumbler Ridge, there was no other option.

Now, in his hand, he held a plane ticket so that he could go see her. He decided to use it on her birthday, which happened to be one day before Valentine's Day.

Gilles has had a rough life. Almost killed in the navy, he's also seen a lot of heartbreak closer to home. One son committed suicide due to drug use. Another was hit by a car when he was only six years old and died. But according to Gilles, "We don't have to live there, even if we never forget." Gilles has a strong faith which has helped him through the years. Now his generosity and kind heart are paying dividends.

Thank you to Gilles for his fine example. Thank you to the three nurses who were so thoughtful and kind. It seems that goodness is still alive and well in Tumbler Ridge.

"A Valentine for Gilles" was originally published in **Tumbler Ridgelines** *- February 6, 2020.*

ABOUT GILLES GODDARD

Gilles Goddard was born near Sherbrooke, Quebec in 1935. Always an athletic and determined person, he went on to join the navy before settling down and raising a large family. He spent most of his life as a long-haul truck driver, but dabbled in a number of other occupations as well. Through some very difficult and sometimes heart wrenching circumstances, Gilles has maintained a strong Christian faith. At the time of this writing, he lives in Tumbler Ridge, BC.

ABOUT TRACY KRAUSS

Tracy Krauss is a multi-published novelist, playwright, and artist with several award winning and best selling novels, stage plays, devotionals and children's books in print. Her work strikes a chord with those looking for thought provoking faith based fiction laced with romance, suspense and humor. She holds a B.Ed from the University of Saskatchewan and has lived in many remote and interesting places in Canada's far north. She and her husband currently reside in beautiful Tumbler Ridge, BC where she continues to pursue all of her creative interests.

"fiction on the edge without crossing the line"
https://tracykrauss.com